by Miche Dahl

FRANKLIN WATTS
LONDON • SYDNEY

First published in 1998
This edition 2002

Franklin Watts
96 Leonard Street
London EC2A 4XD

Franklin Watts Australia
56 O'Riordan Street
Alexandria, Sydney
NSW 2015

Original edition published in the United States by Capstone Press
818 North Willow Street, Mankato, Minnesota 56001

ISBN 0 7496 4586 5 (pbk)

Dewey Decimal Classification 621

A CIP catalogue record for this book is available from the British Library

Printed in Belgium

Contents

Words in the text in **bold** type are explained in the Useful words section on page 23.

Machines

Machines are any **tools**
that help people to do work.
A lever is a machine.
A bottle opener is an example of a lever.

Levers

Every lever is made of
a **bar** and a **fulcrum**.
The bar is the part of the lever that moves.
The fulcrum is the part
that does not move. (Sometimes
the fulcrum is called the **pivot**.)
A see-saw is another example of a lever.

Up,
Dr. David Darling

Lifting a load

A pencil and a roll of kitchen paper
can work together as a lever to lift a book.
The pencil is the bar and
the roll of paper is the fulcrum.
The book is the lever's **load**.
The lever makes it easier to lift the load.

Levers in action

When you play on a see-saw
you are also working a lever.
The plank on which you sit is the bar.
The fulcrum is in the middle of the bar.
Because of the fulcrum
you can lift each other
up and down on the see-saw
very easily.

First-class levers

Hammers and see-saws are both levers.
Each has the fulcrum
between the **effort** and the load.
When a lever has its fulcrum
in the middle like this
it is called a first-class lever.

Second-class levers

A wheelbarrow is another kind of lever.
The fulcrum of this lever
is the wheelbarrow's wheel.
The load is in the middle.
It is between the fulcrum and the effort
(the person using the wheelbarrow).
This type of lever is called
a second-class lever.

Third-class levers

A broom is a lever.
The fulcrum is where one hand
holds the handle at the top.
The load is on the ground
where the broom is sweeping up dirt.
The effort is supplied by the other hand
in the middle of the broom.
This is called a third-class lever
because the work is done
between the fulcrum and the load.

Two or more levers together

The handle of a pair of nail clippers
is a lever that pushes the blades together.
The blades form another lever system.
When two or more levers work together
they are called **combination**
(or **compound**) levers.

Levers are everywhere

People use levers every day
to help them in their work.
Levers make work and play easier.
A world without levers
would have no cricket bats,
baseball bats, tennis racquets,
fishing rods or pianos.

Make your own lever

What you need

A bag with handles
Strong string
A table
Several large stones
Broom handle

What you do

1 Put the stones in the bag. Feel its weight.
2 Tie one end of the string around the bag's handles.
3 Tie the other end of the string around the middle of the broom handle.
4 Rest one end of the broom handle on the table.
5 Lift the other end of the broom handle, raising the bag of stones off the floor.

The bag of stones will feel lighter than it did before. The lever has made your work easier. The fulcrum of your lever is the table, the bag of stones is the load. The load is between the fulcrum and the effort (you), so this is a second-class lever.

Useful words

bar The rigid part of a lever that moves and turns

combination levers levers that work together to achieve the same result

compound levers another name for combination levers

effort the energy provided by the person using the lever

fulcrum the part of the lever that does not move

load whatever the lever moves or lifts

pivot another word for fulcrum

tool something a person uses to do a job

Books to read

Dixon, Malcolm and **Smith, Karen**, *Forces and Movement,* Evans 1997

Kelly, John and **Burnie, David**, *Everyday Machines,* Hamlyn 1995

Ollerenshaw, Chris and **Triggs, Pat**, *Levers,* Black 1995

Turvey, Peter, *The X-Ray Picture Book of Everyday Things,* Watts, 1995

Index